Omar KEBOUR

Photovoltaic Generator Performance Study

Omar KEBOUR

Photovoltaic Generator Performance Study

Feeding a Central Pivot Irrigation System

ScienciaScripts

Imprint

Cover image: www.ingimage.com

This book is a translation from the original published under ISBN 978-620-6-72806-1.

Publisher:
Sciencia Scripts
is a trademark of
Dodo Books Indian Ocean Ltd. and OmniScriptum S.R.L publishing group

120 High Road, East Finchley, London, N2 9ED, United Kingdom
Str. Armeneasca 28/1, office 1, Chisinau MD-2012, Republic of Moldova, Europe
Managing Directors: Ieva Konstantinova, Victoria Ursu
info@omniscriptum.com

Printed at: see last page
ISBN: 978-620-8-50972-9

Summary

Pivot boom irrigation technology represents a major advance in modern agriculture, particularly in arid and semi-arid zones. These highly mechanized systems are capable of covering vast cultivated areas while ensuring uniform water distribution, which is crucial for optimizing crop yields. Thanks to this technique, farmers can not only improve irrigation efficiency, but also make substantial savings on the use of water resources. This contributes directly to food security, by guaranteeing sufficient agricultural production to meet the growing needs of the population.

In the context of the high plateaux and southern Algeria, the adoption of these central pivots has led to significant agricultural development. By integrating this technology into their farming practices, farmers have been able to increase production while coping with the challenges posed by often difficult climatic conditions. However, one of the major drawbacks of using these irrigation systems is the need for reliable, unlimited electrical power.

Faced with this challenge, Algeria benefits from exceptional solar potential. With between 1800 and 3000 hours of sunshine a year, the country is ideally positioned to take advantage of renewable energies, particularly solar power. The integration of photovoltaic systems to power irrigation facilities could offer a viable and sustainable solution.

Our study therefore focuses on the analysis and sizing of a photovoltaic generator designed to power a central pivot irrigation system. This system is designed to move the irrigation spans along a circular trajectory, thus guaranteeing optimum coverage of the cultivated area. In addition, the system's performance will be monitored by the photovoltaic panels, enabling precise and efficient water management. By exploring this synergy between irrigation technology, photovoltaic energy and food security, our study aims to provide practical recommendations for maximizing irrigation efficiency while promoting sustainable agriculture in Algeria's sunny regions, thereby contributing to food stability and the well-being of local communities.

Keywords: center pivot irrigation system, photovoltaic system, solar irrigation.

Contents

Introduction

For a long time and increasingly, water saving has been of great importance in research to improve irrigation techniques, but most efforts and investments in irrigation development in many countries have been focused more on developing water resources than on improving water use at plot level.

To achieve this, farmers have opted for a modern method of irrigation called Rotary Pivot, which is designed to make better use of water, particularly in arid zones with large agricultural areas. The fundamental advantages of this type of device are its ease of implementation, the possibility of automatic operation and its performance in terms of water supply homogeneity.

At present, the electrical distribution network is the main source of energy for the irrigation pivot. Operation is limited to irrigation sites where such power lines do not exist, or due to high construction costs. In the absence of such a line, power for the irrigation pivot can be obtained from a generator (combustion generator).

As we know, in both these sources, the energy supplied is due to the use of fossil fuels such as oil, coal, natural gas or nuclear power.

Recent studies and forecasts warn that the massive use of these resources will inevitably lead to their complete depletion. And everyone is globally convinced of the danger of this process for the environment. That's why we need to look for alternative sources of energy.

Renewable energies such as photovoltaics, wind power and hydroelectric power represent an excellent alternative, and are increasingly used today.

This type of energy is not only free and inexhaustible, it's also very clean for the environment. In fact, it's often referred to as "green" energy, as it completely avoids the pollution generated by traditional sources.

The aim of our work is to study and analyze a central pivot irrigation system and the performance of a photovoltaic system, with the aim of moving the spans along a circular trajectory and controlling the system.

The system consists of a solar array made up of several solar panels, a battery bank for energy storage and inverter equipment, which is responsible for converting direct current into alternating current. Providing the centers with this technology has enabled them to offer a complete solution to the farmer, irrigation equipment plus a stand-alone photovoltaic generator plus manual and remote control equipment, all mounted in a prefabricated concrete building with the appropriate dimensions so that the user can operate it comfortably while protecting it from the elements, sand winds and sabotage.

Our dissertation consists of a general introduction, followed by five chapters presented as follows:

- Chapter 1 describes the central pivot irrigation system.
- General information on solar energy and a description of the photovoltaic system are given in Chapter 2.
- In the 3rd chapter, we explain the sizing methodology for the photovoltaic system powering a central pivot irrigation system.
- The sizing of the photovoltaic system is calculated in chapter 4.
- Results and discussion are presented in Chapter 5.

Finally, a conclusion with proposed perspectives for reaching out to students in the years to come, to make sizing an easy task.

Chapter I:
Description of the central pivot irrigation system

Chapter I: Description of the central pivot irrigation system:

1.1 Presentation:

An irrigation pivot is a device used to spray water to irrigate large areas of farmland. This water, which is sprinkled throughout the property, is manufactured according to the same properties as natural rainfall. Irrigation can be fully or partially integrated, depending on the design of the irrigated field.

The pivots must be ideally located to meet the water needs of the entire site. Several tube spans are available. These main elements are attached to the chassis and motorized posts. The pivot rotates around the central unit to ensure water availability even in these situations. Each pivot point is attached to a wheel, as shown in figure I.1; and their number and characteristics depend on the type of soil.

Figure I.1. Irrigation pivot

1.2 Pivot advantages:

They can be summarized as follows:

- High-speed rotation enables high watering frequencies, which is particularly useful in thin soils with low water reserves.
- Irrigation distribution is very good.
- Total control of irrigation water to save water
- Average life is about 15 years

1.3 . Disadvantages of the pivot:

The most important are:

- Not suitable for small farms
- large initial investment
- high price
- Its use leads to the salinity of the earth's surface.

1.4 System components:

1.4.1 The central element:

This is where the water and electricity come in, and the central element is usually fixed to a concrete slab with anchors sealed inside the block, the volume of which depends on the type of machine construction in Figure I.2 and I.3.

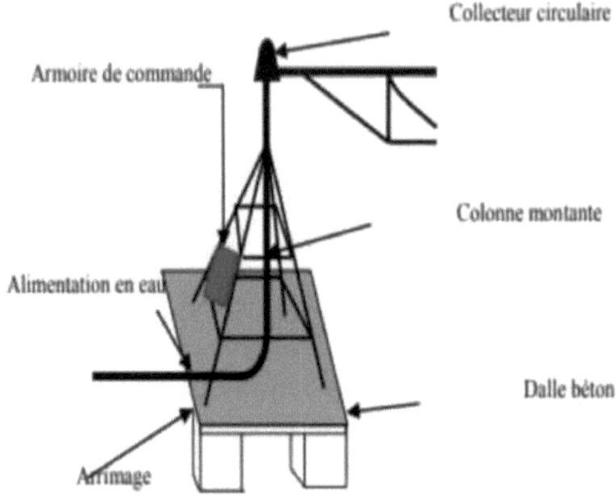

Figure.I.2: Central unit **diagram of an irrigation pivot**

The slab must be able to withstand the reverse moment exerted by the ramp. Its minimum volume depends on the type of structure. Water is directed to the riser, which is the axis of rotation of the unit. The machine is powered either by a circular brush manifold for the entire rotating unit, or by a cord for mains operation.

Figure.I.3: The central element

1.4.2 Mobile towers:

The towers support the pipe. They are equipped with wheels driven by electric or hydraulic (oil-filled) motors controlled sequentially by microswitches to ensure alignment of the spans between the wheels (Figure I.4).

Figure.I.4: Mobile tower of a modern derogation pivot

1.4.3 Spans:

This is between two towers. They act as supports and are made of tubes reinforced by the structure. Ranging in length from 5 to 30 meters, they leave an empty space of 2.5 to 3.5 meters beneath the structure, and have a total height of 3.5 to 5 meters. A large fixed pivot can have more than 15 bays. Mobile solutions are limited to 5 bays. (Figure.I.5)

Figure.I.5: Irrigation ramp span.

1.4.4 Operating principle:

The average Swing Ramp speed is determined by the End Tower execution time. The movement of the entire device is then ensured by successive angle compensators between bays, as described and illustrated below in Figure.I.6.

The angle between two adjacent towers must lie between the two critical angles.

A0 is the release angle (the angle at which the tower in question starts) and A1 is the stop angle (the angle at which the tower in question stops). When the device is switched on (time 0 t), only the last tower is at an angle. When the distance between the tower and tower (n-1) reaches the trip limit, tower (n-1) begins a close run (time 1t), as it has to cover a shorter distance than the tower. Covering the same angle leads to a time (time 2t) when the angle (n) between the towers reaches the stop limit, so there is a gradual ascent towards the pivot axis (Figure I.6).

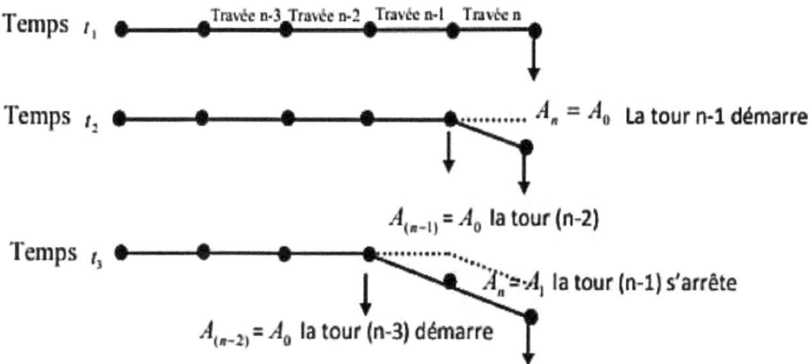

Figure.I.6: Pivot advancement principle

Chapter II:
General information solar energy

Chapter II: General information on solar energy

II.1 Solar energy:

Solar energy is generally used in two different ways:

-One based on electricity production: photovoltaic solar energy.

-The other produces calories: solar thermal energy.

II.2 Photovoltaic solar energy:

Solar or photovoltaic energy converts solar radiation directly into electricity (Figure II.1). To do this, we use photovoltaic modules. These are made up of solar cells connected in series and/or parallel.

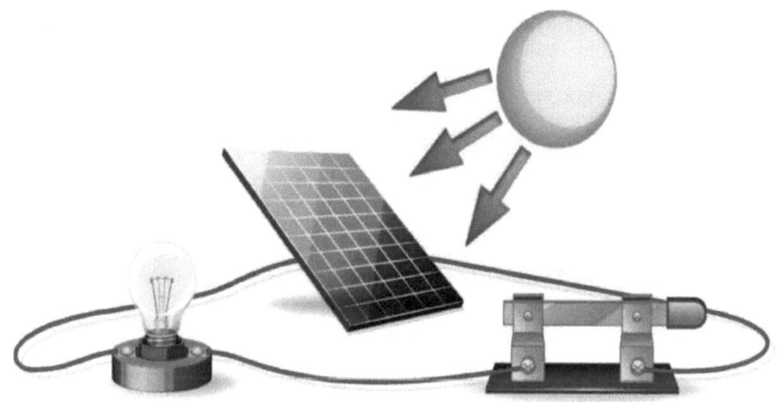

Figure II.1 Photovoltaic energy

II.3. Solar potential in Algeria

Algeria's solar potential is estimated at around 2.6 million terawatt-hours (TWh) per year, or 105 times the world's electricity consumption.

The south is the region that can make the greatest contribution to renewable energy production, given the surface area available and the amount of sunshine. These Wilayas are Tamanrasset, with a potential contribution to the country's solar production of 28%, Adrar 21%, Illizi 14%, and Tindouf, Béchar and Ouargla 7.5% each.

Today, the share of local consumption in national production continues to grow, from 31% to 46% between 1991 and 2017. In Africa, 640 million people regularly have no access to electricity. It should be noted that in 2017, global electricity consumption was 24,800 TWh per year, whereas Algeria consumes 78 TWh per year.

II.4. Photovoltaic cells:

The conversion of solar energy into electrical energy relies on the photoelectric effect, the ability of photons to create charge carriers (electrons and holes) within materials. When the semiconductor is illuminated by radiation of the appropriate wavelength (the photon energy must at least match the material's bandgap), the energy of the absorbed photons enables electronic transitions from the valence band to the conduction band. If the material is polarized, the electron-hole pairs used to transport electricity (photoconduction) can contribute through the material.

However, when a PN junction is irradiated, the electron-hole pairs generated in the space charge region of the junction are immediately separated by the electric field prevailing in this region and drawn into the neutral zone on either side of the junction. (Figure II.2)

When the device is isolated, a potential difference (photons, voltages) is generated across the junction. However, when connected to an external electrical load, a current flow is observed when no voltage is applied to the device. This is the basic principle of solar cells.

THE PHOTOVOLTAIC CELL PRINCIPLE

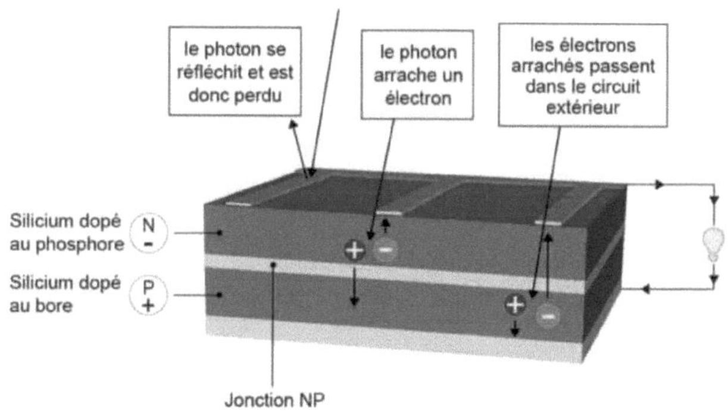

Figure II.2: Operating principle of a solar cell

II.5. Photovoltaic systems:

A PV system generally consists of the above generators connected to one or more elements. Note that photovoltaic systems can be divided into two types:

Those that are self-sufficient and those that are connected to the national grid.

(Figure II.3)

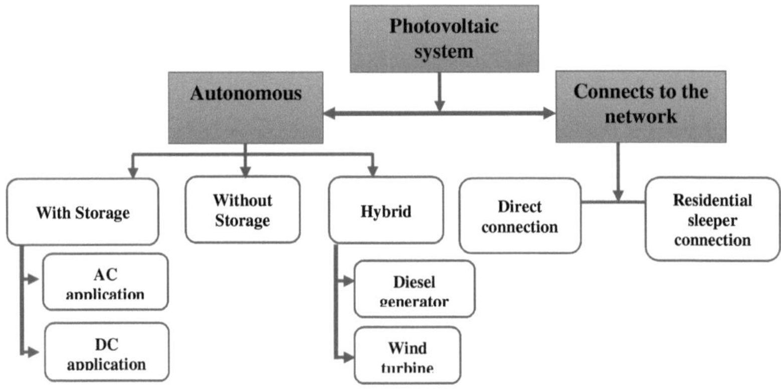

Figure II.3: Classifications of photovoltaic systems.

11.5.1. The Autonomous System:

A PV system is considered autonomous if the load is passive (e.g. lamps, motors, etc.). The air-conditioning system is a battery that stores electrical energy for use when the sun is not shining.

When the sun shines, the photovoltaic generator charges the battery, feeding it directly and storing the energy produced. The PWM (Pulse Width Modulation) and MPPT (Maximum Power Point Tracking) charge controllers protect the battery against overcharging, and the discharge limiter protects the battery against potential deep discharges.

Photovoltaic systems are characterized by their performance and possible applications.

- Stand-alone power supply for consumer products (lamps for garden bollards). (Figure II.4.)

- Building electrification using photovoltaic kits (from hundreds of watts to a few kW).
- Commercial power supply (several tens of watts to several kW).
- Example: Telecommunications.

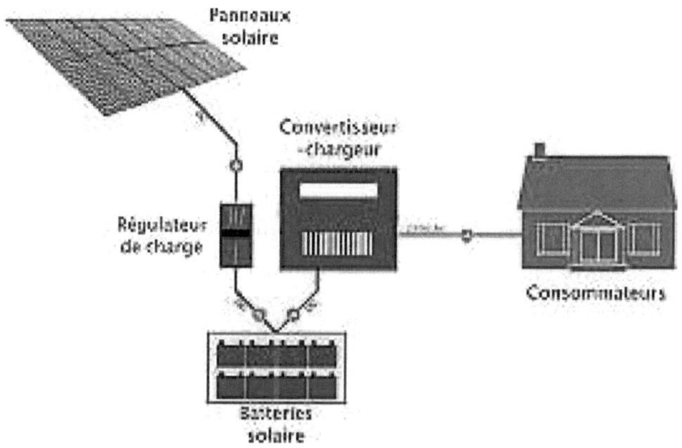

Figure II.4. Stand-alone photovoltaic system

11.5.2. Grid-connected system:

A grid-connected photovoltaic system is one that is directly connected to the electrical grid via an inverter. For grid-connected systems, it is essential to convert the direct current generated by the photovoltaic system into alternating current synchronous with the grid. (Figure II.5)

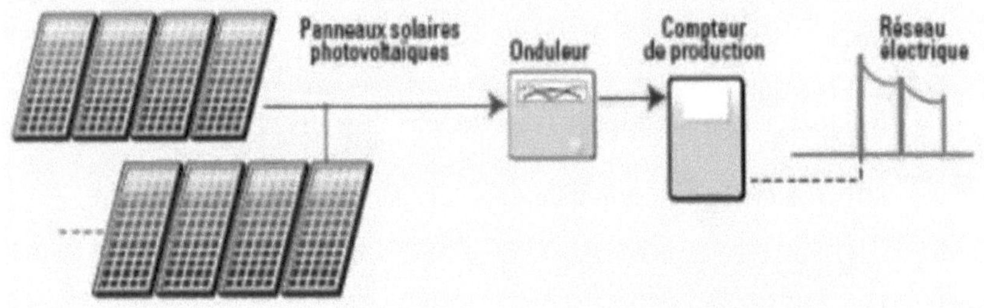

Figure II.5: Schematic diagram of grid-connected photovoltaic system.

11.5.3. Regulators:

The voltage supplied by photovoltaic modules is a variable voltage. This means it varies with temperature and solar radiation. This means that electrical receivers cannot be powered with this voltage. Therefore, to power a device powered from a photovoltaic device, a controller must be used between the powered device and the generator. A controller is a tool for maintaining a quantity of state that corresponds to a setpoint. A regulator converts a fluctuating DC voltage into a constant DC voltage. The charge/discharge controller is connected to the photovoltaic generator and is responsible, among other things, for controlling the battery charging process and limiting its discharge. This function is extremely important, as it directly affects the battery's lifespan. There are several limits, each corresponding to a different type of protection.

Overload, deep discharge, operating temperature, short circuit, etc. New-generation controllers are increasingly sophisticated, offering more advanced functions.

11.5.4. The inverter

An inverter is a device that converts DC voltage into AC voltage. It is used for the following applications in electronic equipment.

Supplies an alternating voltage or current of variable frequency and amplitude. This is the case with inverters used to power AC motors that need to run at variable speeds (the speed depends on the frequency of the current flowing through the machine).

Supply an AC voltage of fixed frequency and amplitude.

Since the energy stored in the power supply's battery is constantly being restored, the inverter has to restore the grid's voltage shape and frequency.

Inverter characteristic for photovoltaic systems:

Inverters for photovoltaic systems are somewhat different from conventional electrical engineering inverters, but the aim of DC-AC conversion is the same.

The main function of a photovoltaic inverter is to find the system's optimum operating point.

A voltage inverter adds voltage to the output in the form of pulse-width modulated (PWM) slots. These slots are ideally suited to powering motors, but are not compatible with the sinusoidal voltages of quasi-sinusoidal power systems (Figure II.6).

Formally, it converts the voltage inverter

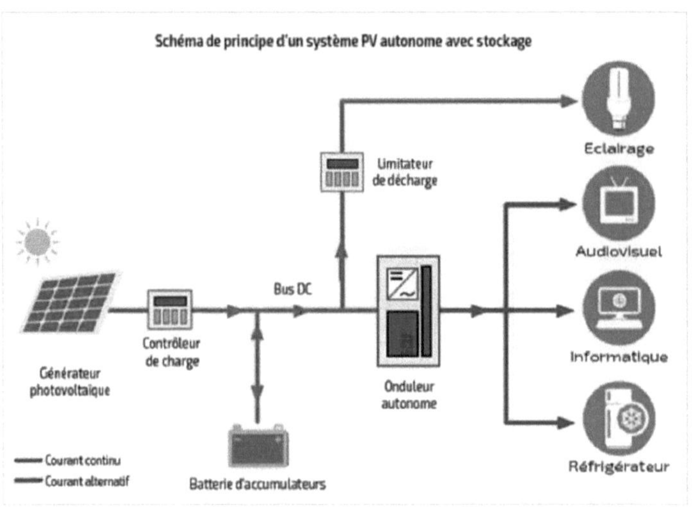

Figure II.6: Stand-alone inverter system

Chapter III:
Methodology

Chapter III: Methodology.

III .1. Methodology:

Sizing a photovoltaic system requires a rigorous methodology to ensure that it meets specific energy requirements. Here is the general methodology used to size a photovoltaic system:

1/ Evaluate energy requirements: Determine the required energy consumption by analyzing the electrical loads to be powered. This may include electrical appliances, tools, cooling systems, etc. Measure energy consumption in kilowatt-hours (kWh) for a given period (per day, per month or per year).

2/ Estimate available sunshine: Obtain data on the average amount of sunshine in your area using reliable sources such as solar databases or weather stations. This data will give you an idea of the amount of solar energy available on your site.

3/ Calculate power requirement: Convert energy consumption into power requirement using the formula: Power required (kW) = Energy consumption (kWh) / Time (hours). This will give you the average power required per hour.

4/ Estimate the size of the photovoltaic system: Using the average insolation, you can estimate the peak power (or nominal power) of the photovoltaic modules needed to meet the power requirement. Divide the required power by the average insolation to obtain an estimate of the number of kilowatt-peaks (kWp) of photovoltaic modules required.

5/ Consider losses and specific conditions: Take into account conversion losses, cabling losses, shading losses, high temperatures, seasonal variations, etc. These losses reduce overall system efficiency. These losses reduce overall system efficiency. Apply a correction factor to account for these losses.

6/ Sizing other system components: In addition to the photovoltaic modules, also size other system components, such as inverters, batteries (if required), cables, protection, etc. This is based on the manufacturer's specifications and applicable standards. This is based on the manufacturer's specifications and applicable standards.

7/ Check profitability and budget constraints: Evaluate the profitability of the system, taking into account investment costs, energy savings and any available incentives or subsidies. Make sure that system sizing is feasible within budgetary constraints.

8/ Perform simulation and analysis: Use photovoltaic simulation software to refine your sizing and evaluate the system's expected performance under different conditions. This will enable you to check whether the system meets your specific energy requirements.

In our case, we used two main parameters to consider: the overall solar irradiation on a collector surface, and the power consumption or energy requirement.

A number of studies have been carried out with the aim of optimizing the sizing of photovoltaic systems. These methods are based on energy balance to determine the storage capacity and output of photovoltaic panels. More recent methods estimate the performance of photovoltaic systems based on the concept of the probability of error in consumption, defined as the ratio between energy deficit and production.

The method we propose consists in establishing energy balances and then calculating the sizing of modules and batteries to guarantee a given level of reliability in terms of consumption. The advantage of this method is that it optimizes the energy consumption of the installation. The difficulty lies in the need to know the hourly irradiation on the installation site for too many years (10 to 20 years). Unfortunately, this data is not often available.

To overcome this problem, we used the "worst month method", since we only have daily solar irradiation values for limited periods. Its principle is to carry out an energy balance under the most unfavorable conditions for the system. In other words, if the system operates in this month, it will operate normally in the other months, thus guaranteeing normal annual operation.

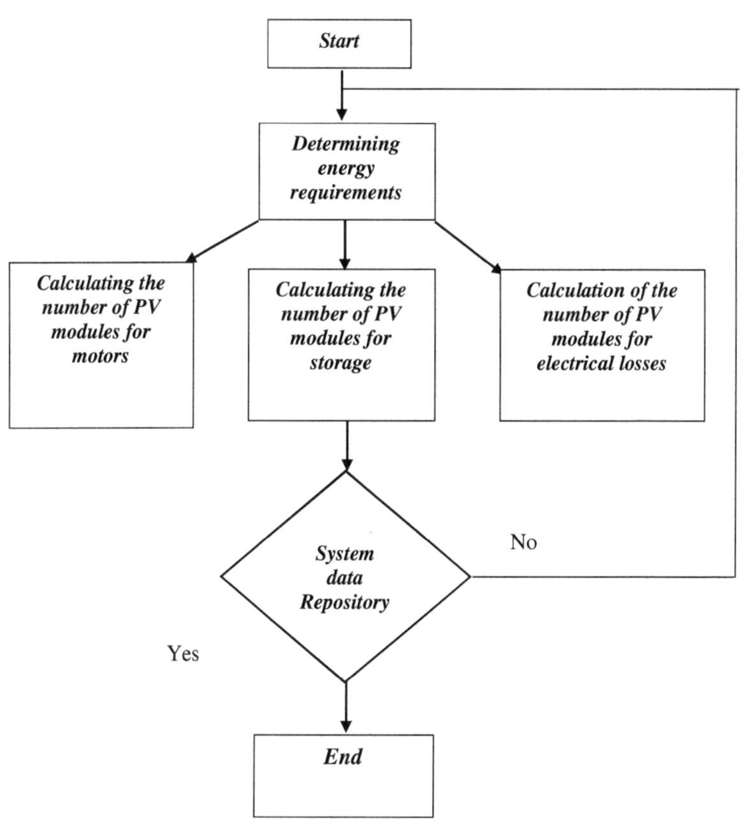
Start
Determining energy requirements
Calculating the number of PV modules for motors
Calculating the number of PV modules for storage
Calculation of the number of PV modules for electrical losses
System data Repository
No
Yes
End

Chapter IV:

System sizing

photovoltaic system

Chapter IV: Photovoltaic system sizing.

IV .1. Estimated consumption

Estimating electricity consumption is based on the energy balance and knowledge of the periodicity of your needs.

Periodicity is in fact the rhythm of electricity consumption. It can be continuous (every day of the year), or periodic (weekends, vacations, periods of the day, etc.).

It's important to find out how much electricity each appliance consumes. Choose appliances with the lowest possible power consumption to reduce electricity costs while maintaining a high level of comfort.

The purpose of sizing is therefore to estimate the size of the various components of the photovoltaic generator system, so as to meet the customer's requirements.

In our case, we're talking about a load of 2 horses (needed to turn one bay of a central irrigation pivot simultaneously).

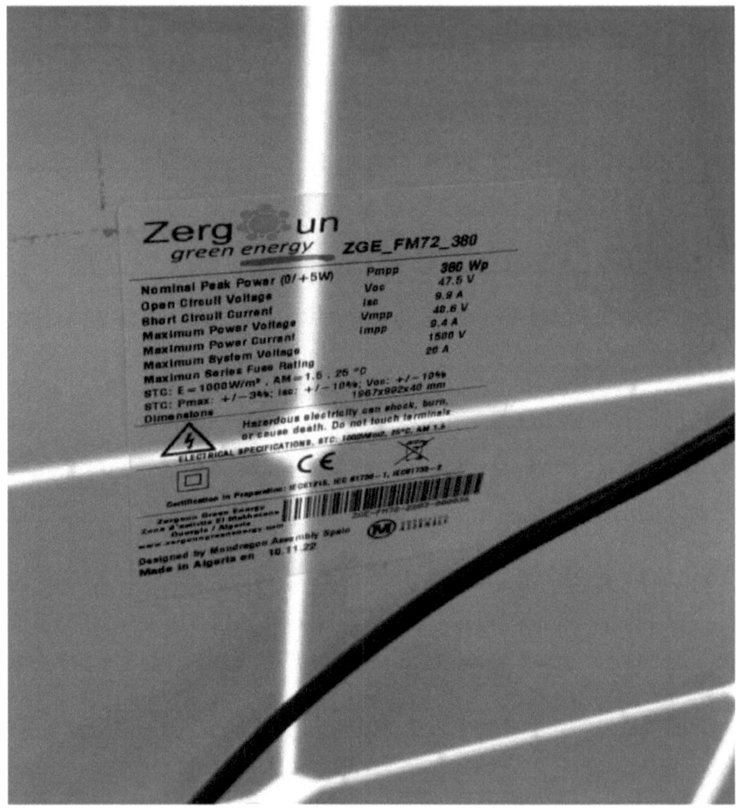

Figure IV.1: Characteristics of the available photovoltaic module

To size a photovoltaic system capable of powering a central pivot irrigation system, we need to calculate the total energy consumption of the motors and estimate the number of panels and batteries required to meet this energy demand. Here are the sizing steps based on the data provided.

IV.2 Calculating the number of modules required for the system

System data

- Motor: 2 hp = 746 Wx2=1472 W , there are 5 of them (but as they don't work simultaneously), so we only count one motor.
- Irrigation time: 10 hours^ Batteries: 2000 Wh capacity, nominal power 250 W
- Battery discharge threshold: 50% of rated power, i.e. 125 W minimum - PV modules: 380 W
- System loss: 10

IV.2.1 Calculating total motor power

Total motor power= 2 cliev; uix746 W=1500 W (rounded)

This means that the total power required to drive the motors is 1500 W.

We need to estimate how long the engines will run each day.

- Let's assume that the engines run from 8:00 pm to 5:00 am, i.e. 8 hours a day (at night).

Daily energy=1500 W8 hours= 12000 Wh

IV.2.2 Calculation of total motor power with losses

The percentage of technical losses in a photovoltaic system depends on various factors, including equipment quality, climatic conditions, site specificities and system configuration. Typically, these technical losses are between 8% and 10% of the photovoltaic system's total capacity.

To cover these losses, we have considered an admissible loss level of 10%.

The power of the two motors is 1500 W, so the power required is:

Adjusted energy (with compensation) =12000 Whx(1+0.10)= 13200 Wh

IV.2.3 Calculating the number of modules required for motors

To power the motors, we need to calculate the number of solar panels needed to produce this energy over the course of a day.

Panel data

- Peak power of each panel: 380 W
- Let's assume an average daily solar output of 10 hours (depending on local insolation).

IV .2.3.1 Calculating the energy produced by a module

Energy per panel per day=380 W* 10 hours=3800 Wh

IV. 2.3.2 Number of modules required

Number of panels= 13200 Wh/ 3800Wh~4 panels;

IV. 4. Calculation of the number of modules required for the batteries:

Each battery has a capacity of 2000 Wh, with a rated output of 250 W. Due to the 50% discharge threshold, each battery can deliver a minimum power of 125 W (i.e. 50% of 250 W).

To supply the 1500 W required by the motor while respecting this threshold, the minimum number of batteries required is:

Nbatteries=1472 W/125 W~11.78N

Let's round up to the nearest whole number, which gives 12 batteries.

In short, we need..:

- **Number of batteries required:** 12
- **Number of PV modules required:** 4

Chapter V:

Results and discussion

Chapter V: Results and discussion

In general, the difference in the number of photovoltaic modules in a photovoltaic system powering a center pivot irrigation mechanism can be attributed to several factors. Here are some technical and economic explanations:

1/ Energy requirements: The size of the photovoltaic system depends on the energy requirements of the center pivot irrigation mechanism. Energy requirements can vary depending on the size of the center pivot, the length of the pivot, the water flow required, the pressure required, etc. If the irrigation mechanism requires a greater amount of energy, a greater number of photovoltaic modules will be needed to meet this demand.

2/ Solar availability: The amount of solar energy available in the region where the photovoltaic system is installed can influence the number of modules required. Regions with less sunshine may require more modules to compensate for reduced energy production. Conversely, in regions with higher levels of sunshine, fewer modules may be needed to meet energy requirements.

3/ Photovoltaic module efficiency: The efficiency of the photovoltaic modules used in the system can also influence the number required. More efficient photovoltaic modules can generate more energy from the same surface area, which can reduce the number of modules needed to achieve the required output.

4/ Budget constraints: Budget constraints can also play a role in determining the number of photovoltaic modules. If the budget allocated to the photovoltaic system is limited, this may limit the number of modules that can be installed, which may result in a lower power output.

In our case, we carried out a detailed analysis of the specific energy requirements of the central pivot irrigation mechanism, as well as the solar conditions, to accurately determine the number of photovoltaic modules needed.

So the number of modules needed for the motors is 4 modules for 1 motor of 2 horsepower equivalent 1500 W.

As far as storage is concerned, after calculation we found 8 modules for optimal charging of the storage batteries for night-time irrigation.

To cover the impact of technical losses due to the Joule effect, estimated at 10%, we need 4 modules, making a total of 16.

However, during our various visits to the farm at El Meniaa, we found that the number of modules in 18

We were able to detect this oversizing of the photovoltaic system with two additional modules, which is important if the number of irrigation points is multiplied. In this activity, the calculations and design of the photovoltaic system must be as optimal as possible, to meet the specific needs of center pivot irrigation, since the cost of the end product (potato-carrot, etc.) depends on the cost of electricity, particularly in the first years of operation after commissioning of the PV system (before amortization).

Conclusion

Outlook

Conclusion Perspectives:

As part of this project, we carried out a study on the sizing of a stand-alone photovoltaic system to supply a central pivot irrigation system installed at a site in the south of the country. On the basis of information obtained during field visits after our technical stay at the farm located in El Menia (wilaya of Ghardaïa), and an estimate of electrical energy consumption and sizing for each element of the photovoltaic system (battery, modules), while taking into account the energy losses that exist.

The sizing of the photovoltaic system and the results obtained and after a comparison with what we found on site during the visit to El Menia, we found that there was an oversizing in the number of photovoltaic modules, a surplus that was not necessary to meet energy needs. Here are a few conclusions we can draw:

Inefficient use of resources: Oversizing of photovoltaic modules indicates inefficient use of resources, as there is excess energy production capacity that is not required to meet the energy needs of irrigation. This can lead to wasted energy and unnecessary costs. Also, poor estimation of energy requirements with the presence of oversizing suggests incorrect estimation or overestimation of irrigation energy needs. It is essential to carry out an accurate assessment of energy requirements in order to correctly size the photovoltaic system. Furthermore. Oversizing has a financial impact, with additional costs associated with the installation and maintenance of unnecessary photovoltaic modules. This can affect the system's return on investment and extend the payback period for initial costs.

The possibility of readjustment: The detection of oversizing offers the opportunity to readjust the system by reducing the number of modules or optimizing their configuration. This allows energy production capacity to be better aligned with actual requirements, which can reduce costs and improve system efficiency. The importance of performance analysis, and the detection of oversizing, underlines the importance of continuous performance analysis of the photovoltaic system. It is essential to regularly monitor and evaluate system performance in order to detect potential problems and make the necessary adjustments to optimize its operation.

Finally, the discovery of oversizing in the number of photovoltaic modules for the system supplying the central pivot for irrigation highlights the need for accurate assessment of energy requirements, optimization of system design and regular monitoring of performance to ensure efficient use of resources and maximize economic benefits.

Perspectives:

The use of photovoltaic systems for rotary center pivot irrigation offers a number of promising prospects. Here are some of the main prospects for this application:

1/ Sustainability and energy autonomy: The photovoltaic system uses a renewable energy source, the sun, to power central pivot irrigation. This reduces dependence on fossil fuels and contributes to environmental sustainability. What's more, the system can be designed to operate autonomously, without the need for connection to the power grid, thus offering energy independence.

2/ Reduced energy costs: Using solar energy to power center pivot irrigation can result in significant savings in energy costs over the long term. The operating and maintenance costs of photovoltaic systems are generally lower than those of conventional systems powered by fossil fuels.

3/ Adaptability to remote areas: Agricultural areas that are remote or far from the power grid can benefit from the use of photovoltaic systems for center pivot irrigation. These systems can be installed in remote locations without the need for costly electrical infrastructure.

4/ Efficient use of farmland: Solar panels can be installed on farmland without affecting food production. By using the space available on agricultural fields for the installation of photovoltaic panels, farmers can maximize the use of their land while producing clean energy for irrigation.

5/ Reduced carbon footprint: Using photovoltaic systems for center pivot irrigation helps reduce greenhouse gas emissions. By avoiding the use of fossil fuels, we reduce the environmental impact and carbon footprint associated with agricultural irrigation.

References bibliographical

References

- Angel Cid Pastor "conception et réalisation de modules photovoltaïques", PhD thesis, Institut National des Sciences Appliquées de Toulouse, France, September 29, 2006.

- C. Brouwer "Méthodes d'irrigation, gestion des eaux en irrigation" Training manual no. 5 1990.

- Evan Derdall "Best Management Practices of a Solar Powered Mini-Pivot for Irrigation of High Value Crops" Master of Science University of Saskatchewan Saskatoon, Saskatchewan 2008.

- Jean Dunglas "Les techniques d'irrigation, Membre de l'Académie d'agriculture" de France Manuscript published in February 2014.

- Jean-François Reynaud "Recherches d'optimums d'énergies pour charge/décharge d'une batterie à technologie avancée dédiée à des applications photovoltaïques", PhD thesis from the University of Toulouse 4 January 2011.

- KY Thierry S. Maurice "Système Photovoltaïque Dimensionnement pour pompage d'eau pour une irrigation gout-à-gout", DEA en Physique Appliquée Option: Semi-conducteurs ANNEE UNIVERSITAIRE 2005-2006.

- léopold rielle and brunon molle ".le pivot", Éditions Cemagref, 1995.

- M. Djarallah "Contribution à l'étude des systèmes photovoltaïques résidentiels couplés au réseau électrique" PhD thesis, University of Batna, Algeria, 2008.

- Petibon, Stéphane. New distributed energy management and conversion architectures for photovoltaic applications. Diss. Université Paul Sabatier-Toulouse III, 2009.

- Peyvieux, Eric. Analysis of the mechanical behavior and shape optimization of an irrigation pivot span. Diss. Bordeaux 1, 1997.

- R. Chenni, "Etude technico-économique d'un système de pompage photovoltaïque dans un village solaire," Universite Mentouri De Constantine 2007.

- Salma Fateh "Modélisation d'un système multi générateurs photovoltaïques interconnectés au réseau électrique" Magistère dissertation université ferhat abbas - setif 2011.

Printed by Books on Demand GmbH, Norderstedt / Germany